THE Best Bug Parade

by Stuart J. Murphy · illustrated by Holly Keller

HarperCollins*Publishers*

LEVEL 1

To Nancy—who makes my life a parade
—S.J.M.

For Lily Lawrence
—H.K.

The illustrations in this book were done with pen and ink, watercolor, and pastel on Rives BFK.

For more information about the MathStart series, please write to
HarperCollins Children's Books, 10 East 53rd Street, New York, NY 10022.

Bugs incorporated in the MathStart series design were painted by Jon Buller.

Library of Congress Cataloging-in-Publication Data
Murphy, Stuart J., date
The best bug parade / by Stuart J. Murphy ; illustrated by Holly Keller.
p. cm. (MathStart. Level 1)
Summary: A variety of different bugs compare their relative sizes while going on parade.
ISBN 0-06-025871-3. — ISBN 0-06-025872-1 (lib. bdg.)
ISBN 0-06-446700-7 (pbk.)
[1. Insects—Fiction. 2. Size—Fiction. 3. Parades—Fiction.
4. Stories in rhyme.] I. Keller, Holly, ill. II. Title. III. Series.
PZ8.3.M935Be 1996 94-49316
[E]—dc20 CIP
AC

Typography by Elynn Cohen
2 3 4 5 6 7 8 9 10
❖
First Edition

THE Best Bug Parade

I am big.

I am bigger than you are.

I am the biggest bug by far.

Bug Parade
Big.
Bigger!
Biggest!!

Good!
Bug Parade

I am small.

I am smaller than small.

I am the smallest bug of them all.

Small.
Smaller!
Smallest!!

Bug Parade
Better!

I am long.
Bug Parade
I am longer than you.

Bug Parade

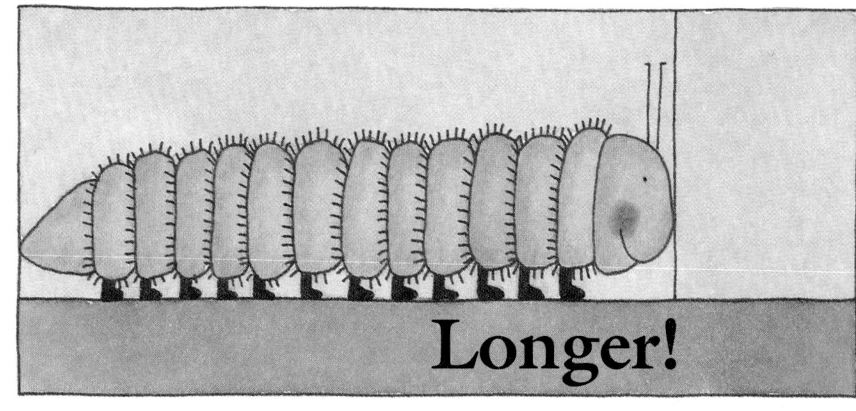

But I am the
longest bug
in view.

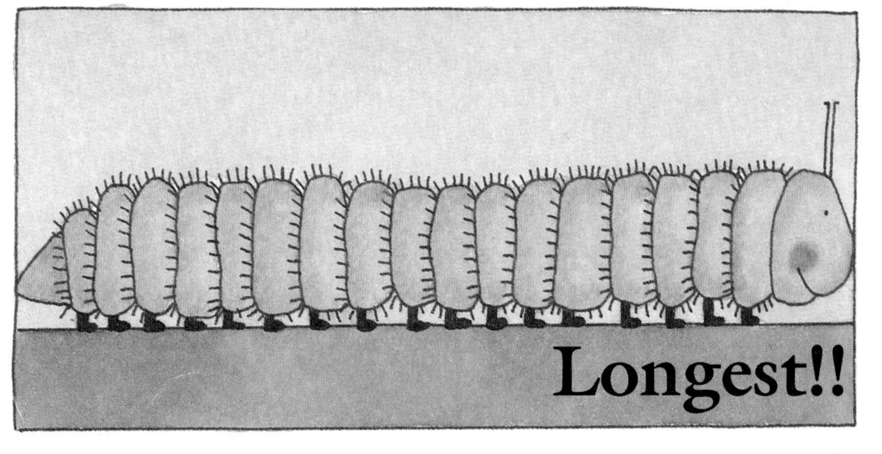

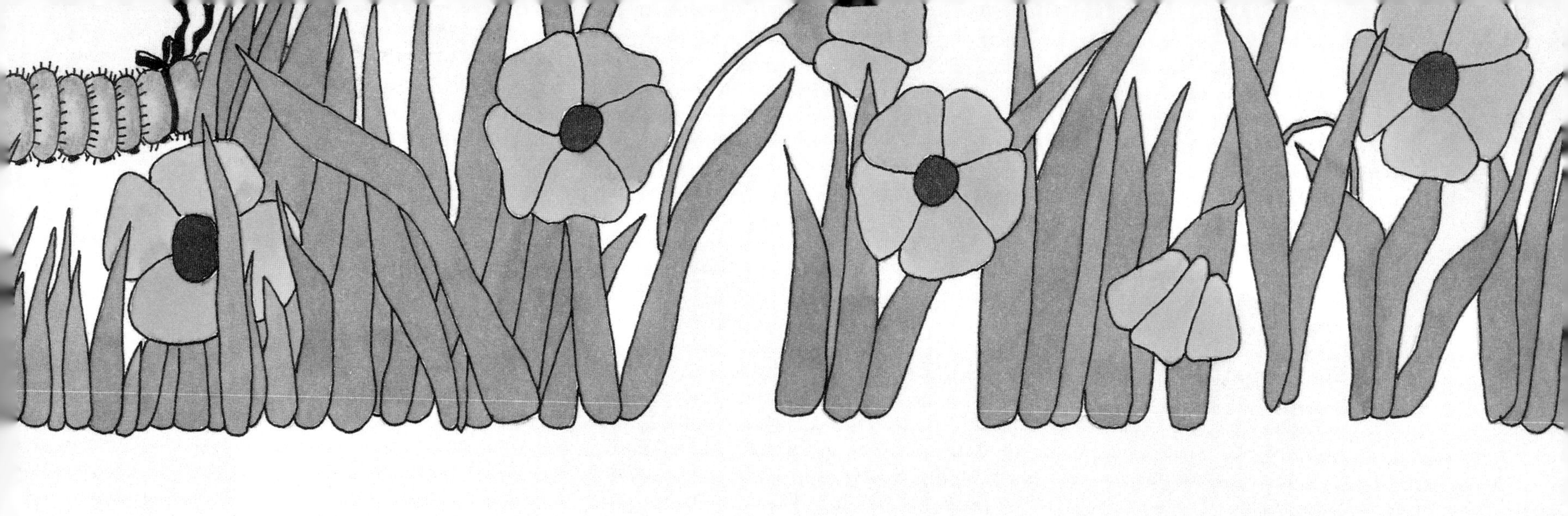

BEST!
Bug Parade

I am short.

I am shorter than short.

I am the shortest bug of this sort.

Short.
Shorter!
Shortest!!

Good!
Better!
BEST!
Bug Parade

When we are all together,
long and short, big and small,

we are not just good or better,

we are . . .

FOR ADULTS AND KIDS

If you would like to have fun with the math concepts presented in *The Best Bug Parade*, here are a few suggestions:

- Read the story with the child and describe what is going on in each picture. Ask questions throughout the story, such as "Do the bugs look the same or do they look different?" and "How do they look different?"
- Together, draw and color some of your own imaginary bugs. Then cut them out and help the child arrange them in order of size. Line them up for your own best bug parade.
- Gather some toys—such as cars, blocks, dolls, or teddy bears—and ask the child to arrange them in order of size. Discuss them with the child using the vocabulary from the book. For example: "Which toy is longer?" "Which is longest?" You can even make another parade.
- Look at things in the real world—family members, pets, furniture, plates, flowers—and discuss their size relationships. "Who is bigger?" "Which is smallest?" Extend the concept by asking such questions as "Which is older?" "Which is youngest?" "Which is darker?" "Which is lightest?"

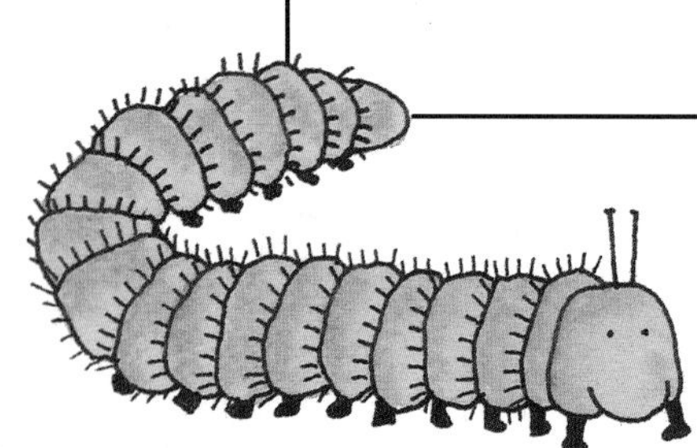

Following are some activities that will help you extend the concepts presented in *The Best Bug Parade* into a child's everyday life.

Cooking: Compare measuring spoons, cups, and bowls while cooking together.

Nature: Take a tape measure or ruler and compare plants in your garden or a nearby park.

Games: Use the vocabulary "Long—Longer—Longest" and "Short—Shorter—Shortest" to discuss the distance a ball is kicked or thrown. Play penny toss on the sidewalk and measure each toss to prove which is longer, shortest, etc.

The following books include some of the same concepts that are presented in *The Best Bug Parade*:

- IS IT LARGER? IS IT SMALLER? by Tana Hoban
- TITCH by Pat Hutchins
- SUPER SUPER SUPERWORDS by Bruce McMillan

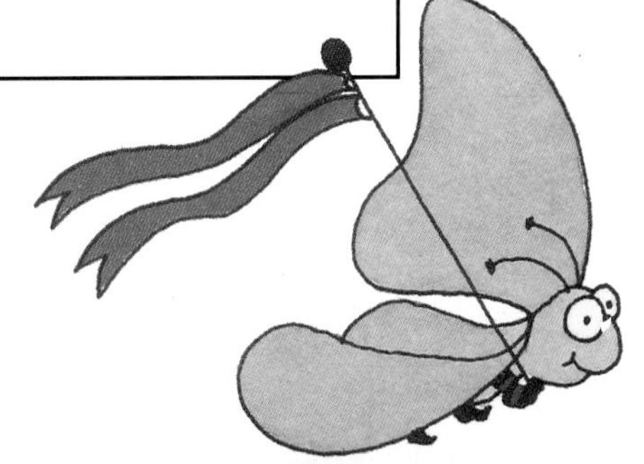